D1161519

PLANTS THAT

SIMON AND SCHUSTER · NEW YORK

GROW ON AIR

JACK KRAMER

Copyright © 1975 by Jack Kramer
All rights reserved
including the right of reproduction
in whole or in part in any form
Published by Simon and Schuster
Rockefeller Center, 630 Fifth Avenue
New York, New York 10020

Designed by Eve Metz
Manufactured in the United States of America

1 2 3 4 5 6 7 8 9 10

Library of Congress Cataloging in Publication Data

Kramer, Jack, 1927–
    Plants that grow on air.

    1.   Epiphytes.   2.   House plants.   I.   Title.
SB427.8.K7   635.9′65   74–23414
ISBN 0–671–21955–3

Contents

CONTENTS

List of Drawings

PLANTS THAT GROW ON AIR

Christmas Cactus.

1·Air Plants

Most houseplants are grown in soil in a container. This is the usual way to grow plants indoors and, although most people believe it is the only way to grow *any* plant, this is not so. As an example, I am often asked why Boston and staghorn ferns do not grow well indoors. Or why the Christmas cactus (a gift from someone's grandmother) does not thrive but seems instead to be dying. One of the reasons why these particular plants do not flourish is that they are epiphytes, or air plants, and do not need nor want to be grown in confined containers with soil. It is not the way they grow in nature, and thus, when they are confronted with different conditions in cultivation— soil in a pot—they soon perish.

Plants differ in their requirements, but through the years, either through misinformation or habit, any and every plant has been immediately potted in soil. Now, as we become more sophisticated in gardening know-how, we realize that different plants have different needs, and that, as far as possible, indoor culture can and should approximate natural habitat. The plants I talk about in this book are those that generally do not want to be grown in soil; indeed, it can mean a slow death for them.

They are air plants, or epiphytes, and, as you will see, include hundreds of lovely, easy-to-grow indoor subjects.

To grow air plants, one must know something about where they grow naturally. Although it is impossible to simulate exactly these conditions at home, many of the crucial elements for these plants can be supplied. And you have one thing in your favor when dealing with epiphytes—by nature, they are perhaps the most adaptable of all plants that have been brought indoors to grow!

Epiphytes grow in all parts of the world, from Central America to South America, with perhaps the greatest number being from Ceylon, Cambodia, New Guinea, and other geographically related areas. In their native lands, they perch on trees, in rock crevices, on poles, telephone wires, and even on other plants—wherever they can find a suitable support. They derive no food from their hosts, and thus they are not parasites.

How do air plants live without soil? Why do they grow in this extraordinary way? The epiphytic development of some plants through necessity is sometimes amazing. In their evolutionary process, such air plants as orchids, bromeliads, and ferns have literally taken to the trees. The forest floor is crowded, with each plant vying with the other for space and light. Through the centuries, some orchids and bromeliads have grown out of the forest floor and adapted to arboreal life, where there is more air and light. These and similar plants are the epiphytes—the subject of this book.

Nature has provided air plants with extraordinary means and methods of survival over a period of thousands of years. To cope with their conditions, epiphytic plants have highly developed root systems and leaves. They absorb nutrients and moisture through their leaves and roots; for example, orchids and philodendrons (though the latter are not true epiphytes) have long aerial roots, and Boston ferns, if you look closely, have brown aerial roots.

Since most epiphytes live in areas with sharply defined

Orchids and bromeliads of many kinds decorate the author's plant room; most of them are grown on tree-fern slabs suspended from trays or ceilings (photo by author).

seasons, they learn how to conserve moisture. Most of them eventually develop thick leaves with a heavy epidermis, thus reducing the loss of water through transpiration. In further specialization, many orchids have thickened pseudobulbs, which act as water reservoirs, and bromeliads have foliage that is usually arranged in a compact, rosette structure that both collects as much water as possible and holds it in storage for the plant's use.

There are remarkable diversifications in plant life, and epiphytes are an incredible example of them. Ferns are basically shade plants residing in moist situations, but like orchids and bromeliads, they do not need soil around the roots to survive. *Polypodium vulgare,* a popular indoor fern, prefers to grow on cork bark, as do the staghorn ferns, Platyceriums. An air plant commonly sold at florists is the resurrection fern, *Polypodium polypodioides.* It has the ability to wither in adverse situations, but starts growth immediately when moisture and light are available, which is why it is commercially sold as a plant that grows without soil, sometimes unfortunately labeled "the miracle plant"—a sensational tag that is not quite necessary. In the case of the resurrection fern, nature has modified the plant to cope with its environment.

However, the resurrection fern is an exception; most epiphytic ferns—and this includes the popular Boston fern (Nephrolepis)—grow in dense shade, on rotted tree bark, or mosses on rocks where moisture is abundant, and yet show little modification in structure.

How I Started

I discovered air plants years ago when I first started growing orchids and bromeliads. Most of these plants are grown in fir bark or osmunda (a type of fern) in a pot rather than in soil. They derive little nutriment from the medium itself; they use it as a support to hold them upright in a container.

One day, in the process of repotting some orchids, I left two of them without containers directly on a redwood bench. I became involved with other things, and when I returned days

*In nature, orchids such as
Telipogon perch high in the trees,
sometimes growing on other
plants. In an arboreal existence
they benefit from good air circulation and bright light
(photo by Paul Hutchinson).*

*Bromeliads, similar in
choice of habitat to
orchids, also abound on
tree limbs in dense forests
(photo by Paul Hutchinson).*

This magnificent specimen of Epidendrum species leads an air plant life in its natural habitat; here perched on a tree limb (photo by Paul Hutchinson).

later to pot the neglected plants, I noticed they were growing as well, if not better, on the redwood slabs. Roots had taken hold, and the plants were thriving. At about the same time, I received a shipment of orchids from Central America, and discovered that many of the plants were mounted on pieces of wood rather than in pots. This seemed more natural, so I put my two plants on some wood slabs. They grew as well as many in pots; indeed, some species prefer this arboreal existence, which is their natural way of growing.

In their native jungles and rain forests, bromeliads grow side by side with orchids high in the trees or on rocks. I started to import bromeliads, a fascinating group of plants, and grew many of them without soil too, sometimes merely setting them in a bowl or appropriate containers in moss or osmunda. Like orchids, numerous bromeliads are epiphytes, and many have a vaselike shape, which enables them to hold water within themselves, like a reservoir.

As time went on, I experimented with other allied plants and discovered the epiphytic cacti—for example, the Christmas cactus—which are forest-dwelling, unlike their desert relatives, and thrive on a diet of moist air.

Air Plants Indoors

All air plants need a suitable support. As in nature, the best way (but not the only way) to grow them indoors is to mount them on pieces of tree branches, wood slabs, compressed tree-fern pieces, moss baskets, and so forth, suspended from ceilings on chain or wire. Epiphytes quite naturally prefer an arboreal existence, where there is plenty of light, air circulation, and humidity. They can be handsome additions in the home when grown on natural materials and they will grow infinitely better this way than if confined to soil in pots. You can also anchor air plants to pieces of wood or rocks in a tray or dish with a base of gravel; the plant rests on the rock and then the arrangement can be used as a table or desk decoration.

This air plant garden indoors is a special accent of the home. Bromeliads grow in tree-fern containers against the wall; a massive orchid grows on the wall in the center and companion plants are philodendrons of several types. There is no soil in the indoor garden (photo by Max Eckert).

Platycerium.

2·Getting Acquainted with Plants That Grow on Air

True epiphytes like many orchids and bromeliads are first choice for air-garden arrangements. However, other plants, such as Philodendrons and Peperomias, can also be grown successfully on cork slabs or pieces of wood. And if you are adventurous, you can try Cordyline, Pothos, Chinese evergreen, or the graceful Hawaiian tree fern (Chapter 7).

Generally, mail-order suppliers are the prime source for epiphytic plants. Only rarely will you find true air plants at local florists or nurseries, and when you do they are apt to be large and expensive. At first, you should choose small growers like the many fine miniature orchids or other small-sized plants.

Finding something to mount plants on can be a delightful treasure hunt. Develop an eye for unusual pieces of wood, shells, small rocks, or a moss basket for your plants; there are an infinite number of natural things to seek out. There are also attractive man-made items that will work (kitchen accessories, such as open-wire vegetable strainers, wire supports, and so forth). Search for beauty in all natural things, but use your imagination: a fruit holder is perfect for plants; a clay pot on its side, suitably arranged in a handsome tray, can be an appealing choice; or you can buy wood slabs and stones (sold at pet shops) or look for driftwood or other natural materials at the seashore or in a forest. Finding the perfect accompaniment for the plant is part of the joy of air plant gardening.

The Best Plants

As mentioned, orchids and bromeliads lead the list of the best air plants. Within the orchid family there are hundreds of delightful miniatures that bear exquisite colorful flowers in their season: spring, summer, or fall. Besides seasonal bloom, orchids also offer a variety of leaf shapes and types, from the grasslike foliage of *Masdavallia igneum* to the leathery leaves of *Gomesa crispa.*

Bromeliads, members of the pineapple family, are fast becoming popular houseplants. Within this group you will find plants with exquisite foliage. Cryptanthus species have bronze and gold leaves reticulated with ivory and red veining; *C. bromeliodes tricolor,* for example, is a tapestry of pink, dark green, and almost black coloring. Concentrate on these leaf colors to make stunning arrangements. Tillandsias are other handsome plants; they have a bottle shape and a gray-green or stark silver color and make handsome dish gardens.

Ferns are emerald or apple-green in color, and their graceful lines make them desirable decorative houseplants. You can try almost any fern if you are adventurous, but the plants that do best on bark or cork are Platyceriums—the staghorn ferns (there are some twenty species)—and Polypodiums.

The cactus family offers some very popular and beautiful candidates, including the well-known Christmas, Easter, and Thanksgiving cacti. There is also a group of true air plants called Rhipsalis: these have cascades of pencil-thin emerald green leaves adorned with white or blue berries in winter.

Philodendrons too, epiphytic by nature, are mainly vining

A large bromeliad is grown in stones near a large rock and flourishes; other air plants (philodendrons) festoon the walls in this unique indoor garden (photo by Max Eckert).

A handsome piece of driftwood is home for three bromeliads, a small *Tillandsia* near the bottom left, an *Aechmea* in center, and *Nidularium*, right. Embedded in osmunda, the plants are wired in place. Here the arrangement is used as a decorative temporary piece, returned to the plant room for watering once a week (*photo by author*).

plants that grow on tree trunks in their natural habitat; they use the tree as a host. If you have trouble growing philodendrons in pots of soil, try them on wood or stone so their aerial roots can attach to the host and thus have a constant supply of moisture. The result will be lush instead of puny plants. Peperomias, especially *P. obtusifolia* and *P. glabella,* are true epiphytes that adapt well to air-growing indoors. Thus there is a galaxy of plants to grow, and there are many different ways to grow them as houseplants "on air," which adds a special touch of glamour to any room. And the plants will grow better than if they were in soil in a container. An unusual way to grow small ferns is to place them on a wire column stuffed with moss; this is a distinctive touch of indoor greenery. Wall panels of epiphytes (and some succulents too) make stunning displays, natural pictures to add grace to a room. Or consider a decorative bird cage, with the wires cushioned with sphagnum moss, holding vining philodendrons to create a living sculpture. Bro-meliads and tiny orchids resting in slatted wooden baskets or fixed to wood slabs are other possibilities and air-garden dish landscapes are certainly different and beautiful. There is no end to the wonders you can create with air plants to make your home colorful and to provide natural plant accessories.

Philodendrons of the Arum family are not all true epiphytes, and yet their habits are akin to orchids because they grow thick aerial roots that are always seeking moisture. In cultivation, many philodendrons are grown on totem poles (described in Chapter 3) so the roots will stay permanently in contact with the moistened bark or tree-fern fiber. Once a philodendron has started to grow, you can cut away the bottom of the plant (as an experiment of course): the rest of the plant will continue to grow—and well—on a moist slab of wood or bark.

A slab of bark is used to support the unusual leafless orchid, the pot merely as an anchor for the wood. Blooming without foliage, Microcoelia is surely a curiosity (photo by author).

Wood is again the host support for this bromeliad tree set in plaster in a decorative dish. The plants are small bromeliads (photo by author).

Beginner's List

Here is a beginner's list of epiphytes to try (see Chapter 6 for more air plants):

Brassavola glauca
Bulbophyllum lobbii
Cirrhopetalum cumingii
Guzmania lingulata
Philodendron bipinnatifidum
P. soderoi
Tillandsia ionanthe
T. juncea
Vriesea malzinei
Zygocactus (many varieties)

3·Air Plants
and Their Needs

Growing plants indoors out of pots may at first appear uncon-
ventional, but there is a great advantage in not having to fuss
with soil or guess about watering schedules. Misting is an easy
process that can be done in a few minutes, and there is never
any possibility of overwatering. (More indoor plants are probably killed by overwatering than by any other reason.) And
soil insects, which many times come in with a potted plant,
are another risk factor virtually eliminated by growing air plants.
    Basically, air plant culture is easy compared to caring for
ordinary potted plants. The epiphytes need adequate moisture
and occasional fertilizing; a source of air (but not drafts); bright
light (but not necessarily sun); and good humidity. A south,
west, or east exposure will generally furnish adequate light
for all air plants, and frequent misting (at least once a day,
but more often in very hot weather) will supply the buoyant
atmosphere these plants need to thrive.
    During the day, average home temperatures of 70° to 75° F
are fine for most plants. At night a drop of ten degrees in
temperature (as in epiphytes' native environment) is very bene-
ficial—indeed, it is necessary to maintain good plant health.
More on culture and care is described later in this chapter.

*Osmunda on a wood
plaque is the base
for a staghorn fern,*
Platycerium hillii.
*The osmunda is wired
to the plaque and then
the plant is wired
in place (photo courtesy
Alberts & Merkel).*

Osmunda and Sphagnum

These natural materials deserve a section of their own because they are vastly important in air plant gardening. Osmunda, the fibrous aerial roots of two types of osmunda fern, is vital to air plant gardening because it provides the cushion or growing medium for the plants. The material resembles spongy wire and holds water, dries out slowly over a long period, and has spaces between the fiber to permit circulation of air. As the material decays (it has no odor), usually in two to three years, it releases mineral nutrients for plant growth. Osmunda is available in large black or brown chunks, generally about a foot square. You can easily cut osmunda with a knife or pull it apart by hand. You might want to soak it the night before using for easier handling. In use, it is attached with wire to the bark, stone, or whatever your plant support, and then the plant is placed against it.

You may have trouble finding osmunda at regular nursery suppliers, but it is always available from orchid suppliers in hobby sacks or by the bale. It is inexpensive, and because it is steamed, there is little chance of potential insect infestation. Osmunda bears no resemblance to peat moss, but nursery shops may try to sell you peat moss as osmunda, so be wary.

Sphagnum is derived from a genus of mosses found in all countries of the cold temperate zone. Green or brown dry sphagnum moss is available in sacks from nurseries. When you ask for this material, specify *live* sphagnum because there is also an artificial type that is not satisfactory for plants. It is a light elastic material somewhat like osmunda but lighter in weight. Sphagnum absorbs water readily and retains it for a considerable time and thus is a suitable mounting bed for air plants. When kept moist it is healthy and green in appearance. Sphagnum is easy to work with and can be used in chunks in the same manner as osmunda. It needs no overnight soaking. Over the years it provides some, but certainly less, nutriment than osmunda for plant growth; however, it is a perfect substitute if you can not find osmunda in your local stores.

Totem Poles, Wood Slabs, Cork Bark Supports

Totem poles are pieces of steamed and shaped compressed wood fiber that are available at most nurseries and generally used for vining philodendrons. These poles can be cut into 10- or 12-inch lengths and used as excellent supports for your air plants. Wire the pieces of totem poles together for larger plants and then insert hooks in the back to secure them to walls. For a novel effect, make a rectangular or square support from poles placed in a dish.

Pieces of wood and outer-bark shells of trees are other good hosts for air plants. (Some orchid suppliers have these materials.) The core is generally removed when you get the shell.

You can also use any type of wooden limb, branch, or driftwood, but select those that have suitable pockets for osmunda fiber. Look for interesting pieces with gnarled shapes.

Compressed Hawaiian tree-fern slabs, also at suppliers, make fine housings for air plants. The slabs come in various

Supports of Various Kinds

A slab of tree-fern fiber holds a small orchid, Oncidium phalaenopsis. *This plant is well established, wires that held it in place have been cut away, and the plant roots have now anchored into the fiber (photo courtesy American Orchid Society).*

shapes and the tree fern is a natural host of air plants, so these make handsome natural containers.

Like bonsai, air plants can also live on small interesting rocks. Look for textured-surfaced pieces about 6 inches or more in diameter.

Recently, costly stones, wood pieces, and branches have appeared at houseplant stores sold as epiphyte supports. To my way of thinking, it is more fun to scout the countryside for your own "containers," but availability is what you are paying for. No matter what support you select—wood, stone, woven reed or rattan baskets—be sure the material is textured so plant roots will have something to adhere to. Seek the unusual, interesting piece that appears natural so that once the plant is attached the complete piece is harmonious and not just a plant on a rock or a piece of wood.

Weathered gnarled wood is excellent as a base for air plants, but it should have a suitable pocket wherein you can lay the plant to attach it to the wood. Smooth-surfaced woods do not work so well. Bark slabs, available at nurseries, are fine; they are generally concave, with natural pockets for plants. Other good supports as previously mentioned are compressed tree-fern fiber slabs (round, square), available at orchid suppliers, or cactus skeletons, which can sometimes be found at florists. These are excellent, gnarled and weathered, and are perfect accompaniments to air plants. Hanging these supports is easy: put suitable hooks in the back of the wood or bark, and attach the supports to walls or hang them from wire supports. (Avoid the manufactured, artificial-looking cork baskets made for terrestrial plants; corklike material—simulated bark—is glued to a wooden pot.)

Moss and tree-fern baskets and allied materials are other fine housings for air plants. These generally have their own hanging devices, so merely attach the plant to the support and allow it to grow into a beautiful green tapestry.

As you can tell, there is much more to growing air plants than just *growing plants;* this is an exciting craft to create unique indoor gardens on wood, stone, and other things. Let the mind dream; use your imagination in creating the "piece."

Wiring and Attachment

At first, until you have mastered them, these simple procedures require patience. It is not enough to slap a plant onto a piece of wood and wrap it with wire. First place a suitable bed of osmunda or sphagnum on the support. The material should be applied like an outer skin to the wood or stone, where the plant will grow. Direct contact is essential. If the osmunda is difficult to break, remember that you can soak it overnight so it is easier to work with. Put layers of it in place against the wood or stone; use thin layers and build up to a 1- to 2-inch bed, depending upon the size of the support you are using. Wire or staple the

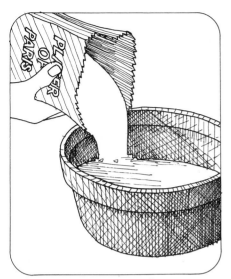

1 Mix plaster of paris and water in planter base

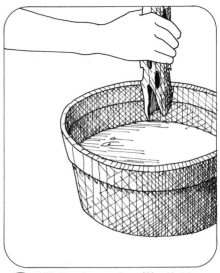

2 While plaster is still pliable insert cactus skeleton

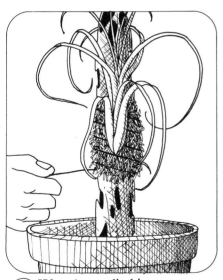

3 Wrap bromeliad in moss and tie with thin wire

4 Cover plaster with sphagnum moss

Bromeliad on a Cactus Skeleton

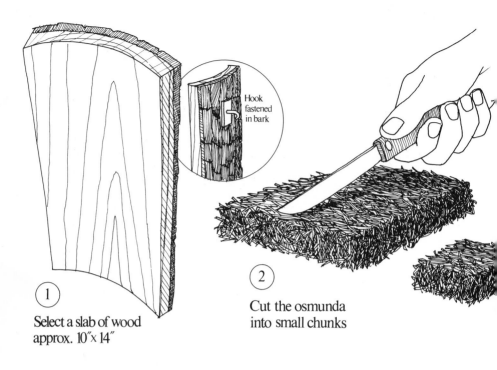

Hook
fastened
in bark

(1)
Select a slab of wood
approx. 10″ x 14″

(2)
Cut the osmunda
into small chunks

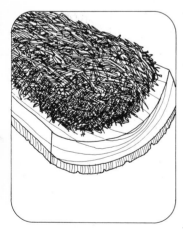

(3) Place a bed of osmunda
1″ to 2″ firmly
against wood

(4) Wire osmunda vertically
to wood with
galvanized wire

(5) Place plant roots into
osmunda. Support
plant with hardware
cloth. Water.

Bark Planting Using Hardware Cloth

Here a small orchid (Gastrochilus) *is grown on a sphagnum support where it blooms yearly* (*photo by author*).

material in place. Cut lengths of wire and wrap them around the osmunda so it is tight against the wood, or use a fine-grade mesh screening and staple or nail it into the support.

To attach the plant, set the crown of the plant (with leaves vertical) at the bottom of the osmunda. If there are roots present, try not to break them; gently insert them into the osmunda or sphagnum. Wire the plant in place with thin galvanized wire. The best procedure is to drill some holes into the back of the wood and then insert wires around the collar of the plant, through the moss and through the holes, and tie them together. The plant should fit tightly in place; it should not wobble or be loose.

If you are using compressed tree-fern fiber in rounds or squares, it is not necessary to use a bed of osmunda or sphagnum. Center the plant(s) depending upon the size of the slab, and then wire it in place by inserting the wire through the slab and tying it in the rear. Tree-fern fiber is porous, so wires will easily pierce the material.

Attaching plants to small rocks takes some experience and must be done with an eye to detail. The plant must occupy a natural indentation in the rock; it must be cradled, not merely set in place. Study the rock first to find the most appropriate hollow. Then lay osmunda or moss on top of the rock, extending it down the sides. Wiring is difficult, if not impossible, with rock, so do the best you can (in time your handiwork will be

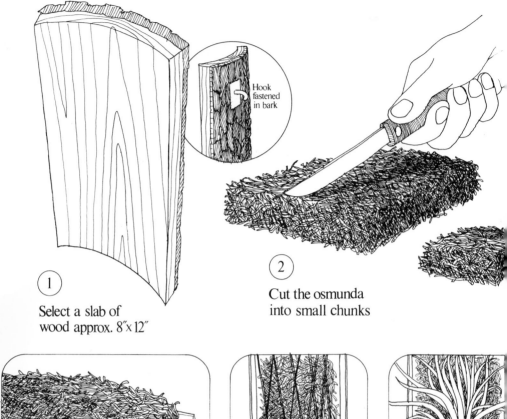

1 Select a slab of wood approx. 8″x 12″

Hook fastened in bark

2 Cut the osmunda into small chunks

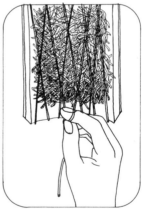

3 Place a bed of osmunda 1″ to 2″ firmly against wood

4 Wire osmunda vertically to wood with galvanized wire

5 Place plant roots in osmunda. Support plant with cross-hatched wire. Wate

Bark Planting Using Wire

covered). To attach the plant, work as you would with the wood.

Preliminary Care

For the first few weeks, observe air plants carefully. If you are not successful, it is because plants are not firmly in place against the support. Keep the air plants in bright light for about ten days—no sun—and be sure the osmunda and wood, stone, or whatever support you are using, is well moistened. Soak the material in the sink once a day, or mist the arrangement at least twice a day. Keep the plants in a warm environment (75° F by day, 10 degrees lower at night). And above all keep humidity high; if you mist several times a day, the humidity will naturally be satisfactory.

After the first few weeks, expose the plant to some sun and put it in a place where there is good air circulation. Keep the roots and osmunda moist, and continue routine spraying with water. In a few weeks you should see fresh white roots; the plant has started its indoor life with you. It is on its own now, and in large part all you have to do is sit back and enjoy. Watering takes only a few seconds a day, and there is no worry about overwatering or underwatering, about whether the plant is growing or not. You can see the roots, which are all you need to examine to determine if plants are prospering. Once the plant is established (six months or more), wires can be cut and removed if you like.

Routine Care

Routine care with air plants involves minimal pruning and clipping and freshening up. Try to keep growth symmetrical, and do not be afraid to cut away any leaves and stems that ruin the total design: the plant will look better and in most cases benefit from the pruning. If the plant does not do well in one area, move it to another spot until you find a place where it grows best. Sometimes an inch or two one way or another can make a world of difference because air might be warmer, cooler, or of better circulation in another location. However, once you find the right place for the plant, then by all means allow it to grow there and do not move it again.

If at all possible, once a month submerge the plant and its support in a bucket of water and allow it to soak a few hours. (You can submerge the entire plant, leaves and all.) This saturates the plant and gives it plenty of moisture for the following days.

There is never any yearly repotting with air plants. In fact, I still have plants on the original tree-fern slabs that they were on some fifteen years ago. However, this is the exception, not the rule. Generally many air plants will outgrow the support in five or six years and should be remounted on new materials.

Because roots are exposed in air plant culture, this affords an easy way to check on plant health. Fresh white roots with green tips mean the plant is thriving; brown roots, except in the case of the brown aerial roots of the Boston fern, indicate the plant is suffering and either needs more moisture or more light. The exceptions are plants that normally rest after flowering; then roots become somewhat shriveled but remain white or occasionally they may brown-off. Even so, it is no cause for alarm. Reduce moisture and allow the plant to rest a few weeks and resume routine watering when you see fresh root growth or leaf growth.

With epiphytes you always know where you are; that is, whether you are caring for them properly. However, there are some tricks. Many epiphytes (especially orchids), as mentioned, go through a dormant period at some time of the year. Then only scant moisture is needed until new growth starts (this may mean a four- to eight-week hiatus). You will be able to determine when this happens because roots become turgid and green.

Feeding

Once plants are established, you can apply some light solutions of plant food. You will need fertilizers that can be sprayed on leaves and plants (these are generally designated as foliar foods). Mix the material with water and then spray it, using an ordinary window-cleaning spray bottle or other efficient mister (many excellent ones are now available).

Use a 10–10–5 fertilizer applied about once every ten days throughout spring and summer. In fall and winter use fertilizer only about once a month. Always use a much weaker solution (add more water) than directions say until you see how plants react to feeding. Some orchids, especially, do not want or need feeding, so let discretion be your guide. Above all, never soak plants with fertilizers to get them growing; it will merely kill them.

If leaf tips start to turn yellow or brown, or if foliage becomes mottled, avoid fertilizing the plants and allow them to rest.

Insect Prevention

For the most part, air plants attract few insects, so you will never have to battle bugs intensively. Sometimes mealybugs or red spiders may attack, but the following chart of causes and controls will help you get rid of bugs:

Condition	Cause	Control
White cottony clusters in stem and leaf axils; undersized foliage; spindly growth	Mealybugs	Spray with Malathion, or remove just a few bugs with an alcohol-dipped Q-Tip.
Leaves deformed, streaked, or silvery, with dark specks	Thrips: Almost invisible yellow, brown, or black sucking insects	Spray with Malathion.
Green, black, red, or pink insects, especially on new growth; sticky, shiny leaves, often cupped; sooty mold; plant stunted	Aphids	Spray with Malathion, nicotine.

PLANTS THAT GROW ON AIR

Condition	Cause	Control
Fine webs at leaf and stem axils and underneath; leaves mottled, turning gray or brown and crumbly	Microscopic red-spider mites	Try vigorous water spray above and below to break webs, or spray with Dimite or Aramite.
Swarming "moths," sooty deposits, leaves stippled or yellowing	White flies	Spray with Malathion.
Clusters of little brown, gray, or white lumps; plant losing vigor	Scale: Insects with hard or soft shells (not to be confused with seed cases on underside of fern fronds)	Pick off, if a few, scrub with strong soapy solution, or spray with Malathion or nicotine.
Holes in leaves	Slugs or snails	Apply Shake-All.
Leaves gray or watery; green or yellow; crown rotting	Bacterial blight	Spray with Captan or Fermate.
Flowers, leaves, and stalks spotted or circled	Virus disease	Best to destroy plant, though spraying every 3 days with Zineb *may* check this.
Leaves coated white	Mildew	Spray with Karathane or Phaltan, or dust with Zineb or sulfur.
Gray mold on flowers and leaves	Botrytis blight	Get rid of badly infected plant; try spraying with Ferbam or Zineb; cut out diseased sections; water less; avoid crowding.

In a greenhouse situation, this philodendron on a totem pole gets good humidity and bright light. As a result it flourishes (photo by Matthew Barr).

4·Decorative Arrangements with Air Plants

Wood and stone supports are generally used for air plants, but there are other inventive ways to have these exotics at home. For example, a rattan basket sling can become a unique, living garden with air plants when suspended from the ceiling. A wall garden is another ornamental way of having plants at home; this is a distinctive, easily constructed picture-frame assembly —always a focal point of a room.

Fern columns are elegant; many of the small ferns make a column of green that can be effectively used as table decoration. Grown with or without a glass dome, the fern column will cheer gray mornings.

Air plants in slatted baskets full of osmunda, or in moss baskets (sold by nurseries) intertwined with plants, are just two other imaginative ways to use plants indoors. And, of course, small trellises in appropriate dishes will also put air plants on display.

There are also the popular bromeliad and orchid trees; branches of driftwood or cypress are used as a background for the plants. This is a fine way to have plants at home, and the "tree" can be placed almost anywhere for a spot of green.

Finally, dish gardens with air plants can be a way to care-free gardening.

Osmunda and Wooden "Planters"

Osmunda planters (available from suppliers) can also be used to grow epiphytes inside and outside the container itself, suspended from ceilings with wire. These planters come in various sizes and shapes—some cylindrical, others tapered or in shapes of animals. Use chopped osmunda or sphagnum to fill these unique containers. As mentioned, slabs of tree fern or osmunda are other good hosts for air plants and these too are fine to use. With all these planters, keep the entire container moist by spraying it with water every day or dunking it in the sink once every second or third day.

Pieces of wood or bark are also available for plants now and these may be of natural or simulated material. Try to avoid the latter. Most of the bark planters have suitable hooks on the back for hanging or holes at the top so they can be suspended with monofilament wire from ceiling supports. In any case, the hanging garden is a fine addition to a room and will always get you compliments from your guests. Further, air plants in them grow lavishly.

Bird Cages

Bird cages are being seen more and more indoors as decorative room accessories, because their graceful lines and sometimes ornate construction make them highly desirable. You can use almost any wire or wooden bird cage to fashion your indoor garden of small plants. Simply wrap or tie bedding osmunda around some of the wire or wood parts of the cage. Put plants in place at various intervals (use small ones), attaching them with wire to the cage. Try to create a design rather than just putting plants in haphazardly. Use large and small plants to create a visual harmony. In a short time, as plants get established, growth will start, and in a few months the bird-cage greenery will be a living hanging garden. Spray moss and plants frequently at the start until plants become established. Then routine watering (two or three times a week) can be done at a sink.

Wall Gardens

These are really vertical living pictures that add a great deal to a room because they look like living tapestry. The mechanics of making a wall garden are quite simple. Select a suitable board and frame. (Fashion your own or buy ready-made ones at frame shops.) Be sure the frame is a shadow-box type, with a preferable depth of 3 to 4 inches to outline the picture. Apply a 1- to 2-inch bed of osmunda or sphagnum to the board with wires. Now wet this thoroughly, and start inserting small air plants such as Cryptanthus, orchids, and perhaps a few succulents. Try to position the plants to make a design; for example, curving lines of one kind of plant combined with vertical lines of other plants will create a handsome arrangement. Wire the plants in place as described in Chapter 3. Allow the wall garden to grow for a few weeks in a horizontal position until the roots have become attached to the osmunda board. Then hang as you would a picture or tapestry. To water the garden, fill the reservoir (3-inch depth) with water that flows over the roots, to keep plants healthy.

Fern Columns

Elegant fern columns were part of the Victorian indoor garden, and these vertical greeneries are handsome accents, distinctive and one of a kind. They can certainly be used today with small air plants such as Tillandsias and Cryptanthus or miniature orchids, or, of course, simply covered in ferns.

The column is easy to construct. Buy a sheet of wire mesh with ¼- or ½-inch openings at your hardware store. Cut the mesh to a suitable height, say, 12 or 16 inches (use a wire cutter and wear gloves to avoid scratching yourself). Now roll

This attractive wall garden panel contains Echeverias in a decorative arrangement; small orchids and bromeliads can also be used in the same manner to create an unusual tapestry (*photo by author*).

A natural moss basket is used to grow one of the epiphytic cacti, Zygocactus. The plant thrives and blooms yearly (*photo by author*).

A natural indoor garden is created on a tree limb. Cryptanthus, Aechmeas and Tillandsias make up the garden and osmunda is used as the base. This arrangement is usually suspended from the ceiling in the author's plant room, but here was put on a table for photographing (*photo by Clark Photographics*).

Inspired by a fern column, this unique vertical garden is no more than sphagnum moss on a wire cylinder; the plants are tucked in place with wires and include Cryptanthus and Tillandsia bromeliads, and several orchids, Brassavola nodosa, and Epidendrums (*photo by Clark Photographics*).

the wire mesh into a column about 5 to 6 inches in diameter. Attach lengths of wire along the column to keep it together. Select a suitable dish for the column; use a rounded dish or bowl at least 10 or 12 inches in diameter and fill it to within 1 inch of the top with wet cement or plaster of paris. (Packaged cement is available at hardware stores.) While the cement is still loose, insert the wire column firmly into it and hold in place for a few minutes. Once the cement has hardened and the column is stable, fill it with osmunda or sphagnum; pack the material tightly in place. Using small plants, carefully attach the roots between the wires with lengths of straight wire. Work from all sides to create a column of green. Be sure the plant roots are tightly affixed to the column so they can penetrate the osmunda or sphagnum. Keep the material moist. In a few weeks the plants will become established on the column and the decorative piece can be used on a table or desk or wherever there is bright light; sun is not necessary. Always try to keep the material in the column moist so plant roots have plenty of water. You can spray the garden or simply trickle water down the column.

Macrame Slings and Moss Baskets

Macrame slings are sold at most nurseries and suppliers. The slings can be used by themselves for air plants. Use small pads of osmunda or sphagnum around the strings and attach roots of plants. You may wire them in place or use twine. In a few months the roots will take hold and plants will start to grow through the macrame sling, creating a lovely greenery. As mentioned, you can attach plants to the strings with small pieces of galvanized wire, but remove the wire once plants have started to grow.

Trellises

As you have probably realized by now, air plants can be grown on almost any type of open or grid work accessory. As you

Davallia ferns grow on rotted tree bark and sphagnum moss in their native habitat. Here the plant is grown in a similar environment. The container is a tree-fern fiber, potting medium is sphagnum (photo by author).

shop, look for the unusual that can become an air garden. I have often used vegetable strainers (open wire baskets) and various other kitchen wire products, devising and camouflaging them for air plant growing. You are governed only by your imagination. The main thing to remember is that air plant roots will grasp almost any porous, textured surface after a while, so do investigate unique holders for them.

We are all familiar with the wooden trellis; smaller versions can also be used for air plants. Place the plants on osmunda in the corner of the trellis grid, much as you would a vine, and start it into growth. The result is a living wall of plants. Wrought-iron trellises will work just as well.

MATERIALS:

(A) 2 - ¼″ plywood boards
24″x 36″

(B) 2 - strips of wood
molding 36″

(C) 2 - strips of wood
molding 24″

(D) 1 - hardware cloth
27″x 40″

(E) 2 - redwood frames
1″x 4″x 28″

(F) 2 - redwood frames
1″x 4″x 40″

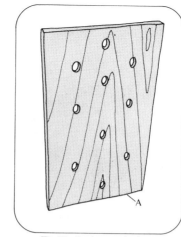

1 Drill drainage holes
in A

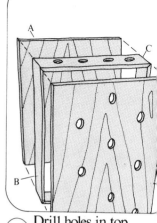

2 Drill holes in top
strip C - assemble
parts A, B, & C

3 Spread sphagnum
moss over frame -
staple D to
frame

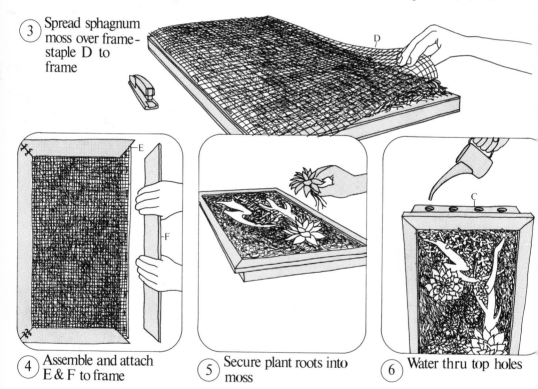

4 Assemble and attach
E & F to frame

5 Secure plant roots into
moss

6 Water thru top holes

Wall Panel

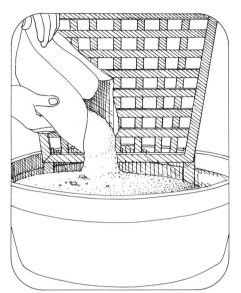

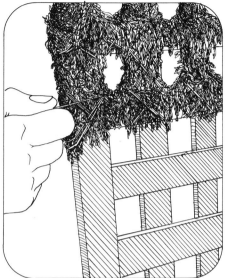

1. Fill pot with plaster-of-paris and cover with small stones

2. Wrap sphagnum around wood with wire

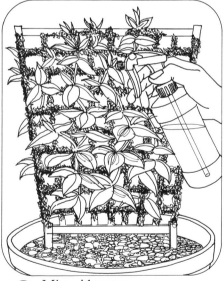

3. Secure plants on to sphagnum with wire

4. Mist with water

Trellis Planting

In this basket, philodendrons and Pothos are used to create an interesting air garden with plants growing from all sides (photo by author).

This kitchen item, originally used to hold fruits, is here transformed into a pyramidal garden. It is stuffed with osmunda and filled with ferns (photo by author).

A lovely dish garden is created in this rock and sphagnum landscape. At the upper left is an Easter cactus variety, in front, small
orchids (*photo by Clark Photographics*).

This is a natural stone the author found. It already had an eroded
pocket and epiphytic Peperomias were used to make an air garden
(*photo by author*).

Dish Garden with Air Plants (cactus and orchids).

Bromeliad and Orchid "Trees"

With bromeliad and orchid "trees" you are trying to imitate nature in miniature. The "tree" should be a gnarled or interesting branch, with several graceful limbs. To start a "tree," center it in a suitable shallow ornamental container or tray; with a plaster of paris mix (sold at hardware stores), position the "tree" and hold it in place by hand until the plaster hardens. Then cover the plaster base with sphagnum or stones.

Now for the planting: Where there are natural curves or pockets, wrap plants in osmunda and carefully attach them to the "tree." Be sure the plants and osmunda are securely anchored to the wood. Place larger plants at the base to give stability to the arrangement, with smaller plants climbing the limbs. Strive for a graceful design, and use different colored plants and different sizes to make an eye-appealing arrangement. Cactus skeletons (available at suppliers) can be planted similarly. General steps to follow for a bromeliad "tree" are:

1. Make a plant pocket in the plaster to hold the tree in the container.

2. Fasten bromeliad or orchids tightly to the sphagnum- or osmunda-wrapped branches. Pull out moss to cover wires.

3. Place pebbles or sphagnum over the plaster base.

4. Use interestingly shaped driftwood, cypress, or fig branches.

5. Spray plant pockets frequently, and once a week water entire tree at the sink.

6. Be sure the arrangement is in bright light.

7. After some months remove wires from sphagnum balls.

8. Use Cryptanthus species, small Guzmanias or varieties of them, and any of the miniature orchids (Chapter 6).

Dish Gardens

Dish gardens are diminutive scenes of nature in suitable containers. A teacup or any bowl can be used, but I find that shallow bonsai dishes with trays are excellent containers. I use

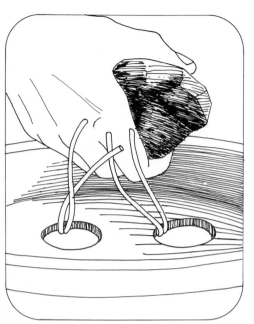

1 Set and wire rocks in place

2 Insert sphagnum moss
   in dish

3 Affix plants to rocks
   in osmunda wads and wire

4 Mist with water

Dish Garden Planting

Like orchids, with long aerial roots to hold on to perches, Pothos and some philodendrons have similar attaching devices: small rootlets, to help them grasp wood or stone (photo by author).

osmunda as a base and then insert rocks of many sizes. Wire the orchids or bromeliads to the stones, and pull moss out to cover the mechanics. Use a combination of small stones and larger ones to create a natural landscape, and fill in with small- and medium-sized plants.

Arrange the dish garden so that large plants are at the rear and smaller ones are in front, in an arc arrangement, so that the front of the container is open for viewing. The front can be filled with colored gravel for accent. Put the dish garden in bright light, and grow compatible plants that like the same conditions, such as Hechtias and Dyckias, Ascocentrums and other miniature orchids.

Once established, a dish garden will make an interesting table piece. Water the plants with a daily spray, and once a week soak the garden at the sink.

Small rootlets on stems can be seen on this Pothos growing on a totem pole (photo by author).

Stone and Shell Gardens

The natural beauty of shells and small rocks blends beautifully with air plants, and these lovely garden pieces can decorate many areas of the home on a table, or a desk. The idea is to select suitable "containers" for the air plants, because not all stones or shells will do.

Look for stones with naturally eroded pockets, so that when they are applied the plants will appear as part of the whole piece, rather than as a tacked-on accessory. You will need plants in scale to the containers. Thus small bromeliads like Cryptanthus and small succulents like Echeverias and Sedums are ideal. First wash the rocks thoroughly, and then insert a bed of sphagnum. Many times, if the rocks have been selected with care and already have eroded pockets, the sphagnum can merely be tucked in place. In other cases you might have to wire the moss bed in place. Now put plants into the planting pockets; do not bury them; place them up to the collar. Use small vine U-shaped clips to keep them in place. Water the natural miniature garden and give it bright light. Twice a week or more, if the atmosphere is dry, soak the small rock garden.

Shells too must be large enough to accommodate plants and, naturally, should be harmonious with the plants. The process for planting remains the same as with small rocks—use a bed of sphagnum, and then place plants accordingly.

5·Air Plants Outdoors

The orchids, bromeliads, ferns, and philodendrons, which do so well as air plants indoors, can also be moved outdoors in summer to patios and yards. Years ago, I summered many air plants on the back porch of my two-flat building in Chicago. Some were placed against the brick wall that made up one side of the porch; they were on wood slabs but soon extended their roots to grasp the cool bricks (and were difficult to remove when autumn came). Others were on hooks along the railing.

Like all outdoor plants, epiphytes benefit from free air circulation and refreshing rains, and indeed, most orchids and bromeliads can be left outdoors until temperatures go below 50° F.

In year-round temperate climates such as southern California and Florida, I have seen air plants as part of the landscape in several gardens. So, if you have a backyard or an open porch, do consider summering some plants there.

On the Patio

The patio or outdoor room is a fine place to summer air plants, and suitable decorative plants can turn the outdoor area into a

Kevin Haapala '73

Brassavola Glauca (orchid).

A fine specimen of
Bromelia balansae *is*
perfect for a hanging
accent at the eave
of this porch.
The plant grows
on a ball of osmunda
(*photo by author*).

Outdoors for the summer,
the author's
bromeliad tree revels in bright
light and even moisture. Plants
are secure in place now and
wires have been removed
(*photo by author*).

tropical haven. So transfer your indoor plants onto posts, roof eaves, or wherever there is a place for a hook. You can have colorful flowers through the summer, and except for occasional watering, you will not have to care for plants at all: nature will do it for you.

If the patio has no vertical supports, or if there is simply no place for plants, you can easily make some wooden plant stands to accommodate small plants. A simple-to-build wooden pyramidal garden, such as the one shown on page 61, works fine. Planting pockets provide places for your plants, and the design of the planter gives a combination of shade and sun so plants are not always in direct sun. Make shelves at least 10 inches deep and 7 to 8 inches wide for maximum use, and construct the planter from redwood. Fill the planters with osmunda and attach plants with wires, as previously explained. Place this garden anywhere, and move it indoors near a window when autumn comes. Provide a suitable tray or mat so you will be able to water the garden.

Wooden posts of random heights can also be used to fashion vertical air gardens. Move the plant and support to the desired position and attach them with a hook at the back into the post. This same procedure also works on walls.

Plants such as Platyceriums and Davallias with their cascading fronds make excellent decoration for outdoors, adding a tropical feeling. The Davallias are airy and charming, while Platyceriums are bold and dramatic. Group several together for summertime enjoyment outdoors. Keep plants on their original supports when you place them on the patio or terrace.

Orchids and bromeliads are fine outdoor decorations too, and can add a festival of color in the outdoor area. If they are growing on cork bark or tree-fern slabs, they are easily moved outdoors on posts or suspended on wire from roof eaves and overhangs. If you know you are going to use these plants outdoors, order summer-flowering kinds, for there are orchids and bromeliads to bloom in all seasons.

Bromeliads (center), staghorn ferns (upper right), and philodendrons (upper left) make up this lush outdoor garden. Note orchids in fiber baskets at upper left (photo by Hort-Pix).

Outdoors, keep epiphytes out of direct sun, but be sure they have bright light. Water plants daily unless there is frequent rain. When weather becomes too cool for plants (generally after Labor Day in most states) move plants back to their original places indoors.

In the Garden

In year-round temperate climates, orchids, bromeliads, and some ferns can be grown all year long on trees. The process is the same as for growing them on a piece of wood indoors: simply apply a bed of osmunda and then attach plants. The best location is where *bright* light, rather than intense sun, reaches plants. Check plants often to be sure they are well anchored to the host support. It takes plants about a year to become established, at which time wires can be cut and discarded; the plant will have anchored itself in place by then. Basically, almost any orchid and most all bromeliads can be grown in this manner. Platycerium ferns and Davallias can also adorn trees outdoors if weather is consistently warm.

Epiphytes will also grow beautifully on almost any plant, whether it is a cactus or a shrub. Position the plants in crotches. Do not fear that this will harm the host plant; the epiphyte will use the anchor as a support but never take nutrients from it. Remember that epiphytes are not parasites; they take their moisture and nutrients from the air, not from the host plant.

Plants can also be attached to porous rocks and will do very well in such a situation. Use rocks that have texture so roots have something to grasp. You will have to wire plants in place but once established you can remove the wiring. By the way, lavarock and featherrock are fine for plants too.

In the garden, on tree limbs or rocks, epiphytes heighten the outdoor scene, for they do impart a tropical feeling and as years go by you will be amazed at how well some orchids and bromeliads—ferns too—will do. And best of all, of course, care

A unique pyramidal wooden garden has planting pockets; filled with osmunda they can be planted with small succulents, orchids, bromeliads, and even ferns, to create a stunning outdoor accent (photo courtesy California Redwood Association).

61

A thriving Echeveria grows on a sphagnum wall panel and adds a beautiful touch to an outdoor porch (photo by author).

is at a minimum as long as there is ample rain. In very warm, dry weather, of course, you will have to supply moisture, but it is a simple matter of gently hosing plants down.

6·The Most Popular Air Plants

Orchids

The Orchid family includes some 35,000 species, from all over the world. The majority are epiphytic plants, growing on tree branches in the forest.

Generally, for indoor use orchids are grown in fir bark (the bark of the Douglas fir tree steamed and chopped) in pots. The bark is used as a support, not as a nutrient medium. The plants can also be grown on osmunda. In this medium they do derive some nourishment, but basically most of their moisture and nutrients come from the air.

Because this is such a large group of plants, it is difficult to suggest the best orchids, so I have recommended the ones I have grown as air plants on bark or cork slabs or tree-fern fiber, and a few that have responded while perched on small rocks in ornamental containers in my garden room. For the most part I grow my orchids in certain conditions; by following these suggestions you too should have success with the plants.

Follow cultural hints included with my descriptions. The conditions are:

1. daytime temperatures of 70° to 85° F, with a 10- to 15-degree drop at night
2. bright light, with a few hours of late afternoon sun (use trellis screens to filter summer sunlight)
3. a range of humidity from 60 to 70 percent
4. a daily misting of plants by hand with a misting bottle
5. water in the winter three times a week and keep plants rather cool (about 55° F) at night.

Orchids are basically arboreal, so use them as hanging plants. Insert S-hooks or wires into the wooden support, and suspend plants from chains or wires anchored to the ceiling. It is essential that orchids have good ventilation from all sides, which is why pot culture is not so successful as air growing. If hanging is out of the question because of water stain on floors or other problems, then lay the tree branch or stone support in a dish, as explained in previous chapters.

As mentioned earlier, most of the epiphytic orchids have water-storing pseudobulbs, so plants really can take care of themselves (but misting is a daily requirement). The leaves of many orchids are tough and leathery, another reason why they can survive with little water (like cacti, which also store their water). Orchid roots are thick and long, and the more roots a plant has the better it will grow. Active roots are green-tipped, but white roots should not be cut. Brown dead roots, however, can be removed without harming the plant. As in nature, roots will clamber all over and attach themselves to whatever is at hand—wood rails, window mullions, even glass panes.

The following plants are the ones I have worked with and grown strictly as epiphytes out of pots:

Aerides crassifolium. Small species, with purple flowers and thick, leathery dark green leaves. Grows well on compressed tree-fern fiber.

A. odoratum. My favorite Aerides. Has bloomed faithfully every summer, with hundreds of heavily scented waxy white blossoms. Rather large plant that grows on a tree branch strung from ceiling wire.

Ascocentrum ampullaceum. Small straplike leaves to 4 inches. Bountiful crop of erect stems of red flowers; needs little care.

A. curvifolium. Small plant to 5 inches, with round magenta flowers.

A. miniatum. Smallest of the group, with dozens of bright orange flowers in spring. Easily grown and a charmer.

Bifrenaria harrisoniae. Dark green 4- to 6-inch leaves, one leaf to a bulb, with beautiful 3-inch flowers. After they bloom, let plants dry out completely for about a month.

B. tyrianthina. Large plant, with solitary leaves that produce reddish-purple exotic flowers. Treat the same way as *B. harrisoniae.*

Brassavola digbyana. Large, to 30 inches, with solitary and mammoth greenish-white flowers. One of the most beautiful orchids; I keep mine on a piece of compressed tree-fern fiber.

B. glauca. Solitary leathery gray-green leaves and large waxy white flowers. Will not succeed readily in pot but grows well on rock or tree branch.

B. nodosa. Good small plant, to 4 inches, for rock growing. Lovely scented blooms.

Brassia maculata. Produces six to twelve flowers, with sepals about 4 inches long. Petals are slightly shorter and greenish-yellow spotted with brown; white lip spreading and marked with brown or purple. This is a good Brassia to start with.

Broughtonia sanguinea. Small 2-inch leaves; handsome red flowers on graceful long stems bloom throughout the year. Excellent, easy-to-grow epiphyte.

Bulbophyllum barbigerum. Narrow greenish-brown sepals and petals; the yellow lip is long with two dark purple markings and tufts of purple threads. A bizarre flower. The slightest breath of air puts these hairs in motion, giving the plant an almost animallike quality. Does very well on tree-fern slabs.

B. lemniscatoides. Remarkable plant, but difficult to locate. The flower spike is drooping, with many tiny dark purple flowers decked with white hairs and carrying white, rose-spotted, ribbonlike appendages.

B. macranthum. Star-shaped flowers about 2 inches across. They are red, splashed with yellow. Good showy species to start with and easy to grow. It has blossomed twice for me since I bought it two years ago.

Catasetum russellianum. Produces flowers 2 inches across, pale green, veined with dark green lines. Plant is very free-flowering and always distinguished by a strong scent of roses.

Cattleya walkeriana. A small plant that bears a magnificent lavender-purple flower. Will require coolness to bloom (62° F at night). Grows well on tree-fern slab [*page 70*].

Cirrhopetalum cumingii. Masterpiece of tiny, vivid, ruby-red flowers arranged in a half circle. Good species to start with and a most interesting plant [*page 70*].

C. longissimum. Produces flowers remarkable for the length of the side sepals, which are sometimes 12 inches long. Flowers are ice-cream pink or buff, with purple lines running to a purple spot. This flowers in winter for me and is spectacular.

C. ornatissimum. Interesting, although not so pretty as others in the family. Flowers are pale purple-brown, about 3 inches long; sepals and petals are tufted with hairs and the lip is red-purple. Has a rather unpleasant odor.

Dendrobium aggregatum. Dwarf plant, about 10 inches high. Produces from the sides of the pseudobulbs pendant spikes covered with small, scented, vivid yellow flowers. This species has flowered for me regularly for the last five years, always in March.

D. pierardii. Produces 2-inch, paper-thin flowers that are blush-white or pink and veined rose-purple. Very dependable plant [*page 70*].

Hexisea bidentata. Bears clusters of bright red flowers, in spring or summer. About 1 inch long.

Masdevallia picta. Medium-sized grassy orchid with 2-inch flowers of yellow-red-purple.

Notylia xyphorous. Sharp, cactuslike, 1-inch leaves. Has tiny, pale purple flowers that appear in late summer for me.

Ornithocephalus bicornis. Most popular species. Height 3 inches. Produces tiny greenish-white flowers.

O. grandiflorus. Bears 5-inch leaves with 1-inch flowers, white with an emerald green splotch at the base of the lip.

Pholidota chinensis. Small creamy-white flowers evenly spaced on a pendant scape. This plant blossoms in late summer for me.

Platyclinis filiformis. Also known as *Dendrochilum filiforme.* Has 5- to 6-inch grassy foliage and pendant chains of perfectly arranged, small yellow flowers in summer.

Pleurothallis barberiana. Bears 1-inch leaves. Produces slender nodding flower spikes crowned with white flowers spotted purple. This has been a dependable flowering orchid for me.

P. grobyi.   Bears small bright yellow flowers, streaked with crimson, on 1- to 3-inch foliage.

P. picta.   Densely tufted species, with 2-inch leaves. Bears yellow (sometimes orange) flowers, half an inch across.

Polystachya luteola.   A plant I grew for many years. Height of 4 inches, with erect spikes bearing several pleasantly scented yellow flowers shaded green.

Rodriguezia decora.   Produces a slightly larger flower, white or rose-spotted red; the spreading lip is usually white. This species does better on a raft or slab of tree fern because of its creeping rhizomes.

Scuticaria steelei.   Light yellow, almost orange, waxen and showy flowers 4 inches across, spotted red.

Sophronitis coccinea.   Produces a bright red flower almost 3 inches across.

Stelis ciliaris.   About 6 inches tall, with many dark maroon or tiny purple flowers. The margin of the sepals is usually fringed with hairs. This species bloomed only once for me; the second year I killed the plant by allowing the osmunda to become water-logged.

S. micrantha.   Height of 6 inches, with small flowers. The sepals are yellow or greenish white; the petals dark red. Usually this blossoms in autumn and is very pretty. I have been unable to locate a mature specimen.

Symphoglossum sanguineum.   Small leafy plant with wands of small reddish flowers. Grows well on wood or tree-fern fiber or in dish-garden arrangements [*page 71*].

Telipogon. A rare genus of orchids; occasionally plants are for sale. They bear exotic triangular-shaped beautiful flowers.

Vanda suavis. Cream-white flowers spotted red-purple. Very free-flowering. There are many hybrids, so flower color varies considerably.

V. teres. Fleshy pencillike foliage. Flowers are pale rose or magenta. This species needs a decided rest in winter.

Bromeliads

Native to tropical America, bromeliads are mainly epiphytic plants that offer great indoor decoration for little effort. Growing side by side with orchids in arboreal existence, these are beautiful multicolored-leaved plants. Although they also produce incredible brightly colored flowers, to my eye it is the foliage that is stellar: some bromeliads have burgundy leaves, others are golden-green or almost yellow, and many are variegated plants, for example, *Neoregelia carolinae tricolor,* with green, white, and pink foliage. Tillandsias, excellent tree or stone plants, are silver-gray or green, and many are small and perfect for air growing.

Because there are almost 1,000 species, bromeliads have an incredible variety in shape, but basically most are vase- or rosette-shaped. Root systems are not so extensive as in orchids, but, as described in Chapter 1, bromeliads are bestowed with a natural miracle: their vase within the rosette holds their own supply of water.

Bromeliads are especially delightful as indoor decoration because they adapt well to all varieties of epiphytic culture and displays.

Like most air plants, bromeliads' main requirements are a good circulation of air (they like high places and perches) and good light (the more light they get the better the leaf color). And as houseplants they are tough to beat, because they will,

Masdevallia (unidentified).   Cirrhopetalum cumingii.

Cattleya walkeriana.

Dendrobium pierardii.

Neoregelia carolinae tricolor.

Symphoglossum
sanguineum.

All photos this page
by author

Tillandsia ionanthe.

Tillandsia cyanea.

if necessary, survive extreme conditions—shade or intense sun.

Within this group, there is a galaxy of plants that you can grow for your air gardens, and just how you use them indoors depends upon your imagination. Most bromeliads will grow on practically any support and need only initial installation to become established. Try any of the following bromeliads:

Aechmea fasciata. Sometimes called the urn plant. A window gardener's delight, with 24-inch leaves. Produces a large pink inflorescence, with hundreds of tiny blue flowers in summer. Fine for the beginner, and sure to bloom.

A. racinae. Small, with light green leaves and yellow and black flowers on a pendant spike. Orange-red berries appear appropriately at Christmastime and they last until April. Try this one in a hanging basket; grow it in light shade.

Billbergia 'Fantasia'. Hybrid. Perfect size for the windowsill, with green leaves spotted ivory, green, or rose. The pendant inflorescence of scarlet bracts and blue flowers makes it most appealing.

B. pyramidalis var. concolor. Favorite with collectors. Broad-leafed, bottle-shaped, golden-green plant. The compact flower head is densely set with pink to red flowers on a short scape. A very beautiful Billbergia that blooms in winter.

B. sanderiana. Makes a medium-sized few-leafed vase. The foliage varies with light exposure, from grayish-green to almost red. The nodding inflorescence is rose, green, and blue.

Bromelia balansae. Has spiny, hooked, dark green leaves 2 to 4 feet long with red bracts and a cone-shaped inflorescence of blush-white flowers. At blooming time the center erupts into a volcano of flame red—a splendid sight.

Catopsis berteroniana. Medium-sized plant. Has apple-green leaves, green bracts, and small white flowers that usually appear in the spring.

C. floribunda. Small and bushy bottle-shaped plant, with bright green foliage and pretty white flowers on a tall arching spike that appears in spring. Excellent for the beginner.

Cryptanthus acaulis. Grows about 5 inches across. Pointed apple-green leaves, with a slight gray overcast. Tiny white flowers hide deep in the center of the plant.

C. beuckeri. Slightly different in the genus, with green and cream spoon-shaped leaves.

C. bivittatus. Sometimes sold as *C. roseus pictus.* Perhaps the most common Cryptanthus. Frequently used in a dish garden because of the spectacular salmon-rose and olive green leaves.

C. zonatus. Small species. Has broad, lacy, brownish-green leaves crossbanded with irregular silver markings. The white flowers are hidden in the leaf axils.

Dyckia brevifolia. Also known as *D. sulphurea.* Grows 10 to 15 inches across, with stiff and succulent dark green leaves. The tall spray of orange flowers appears at various times. Will do wonderfully in the garden, and can take some light frost.

D. fosteriana. The most popular species, and rightly so. Makes a handsome medium-sized plant, with silvery purple rosettes that cascade over the pot, recurve, and form a fountain of leaves. Small orange flowers appear in spring or summer.

D. rariflora. Small, with gray-green foliage and orange flowers.

Guzmania berteroniana. Medium-sized, with leaves to 20 inches. The showy inflorescence has orange-red bracts with yellow flowers. A fine plant for the windowsill.

G. lingulata. Handsome species, about 18 inches across, with a star-shaped orange flower head. *G. lingulata var. splendens* has bronze-rose leaves, and *G. lingulata var. minor* has a red-orange inflorescence with white flowers.

G. musaica. Sure to please houseplant enthusiasts. The leaves are 24 inches long, bright green, banded and overlaid with irregular lines of dark green and wavy purple markings on the reverse. The flower spike is erect and turns red at flowering time. The white waxy flowers are set tight into the poker-shaped flower head.

Hechtia argentea. Has large recurved leaves in a symmetrical rosette with a tall spike of orange flowers. A good showy bromeliad.

Neoregelia carolinae. Perhaps the showiest in the genus. The tapered leaves are dark green; the center of the plant is bright red before blooming. My plant was in full color for nine months. Undoubtedly one of the finest houseplants obtainable [*page 71*].

N. spectabilis. The painted fingernail plant. Has spiny, leathery green leaves tipped cerise. The small blue flowers usually appear in warm weather.

Nidularium innocentii var. striatum. Has creamy white and green striped leaves, a flaming red center, and white flowers in summer. This species is somewhat temperamental and requires warmth.

Tillandsia bulbosa. Has a bulbous base with narrow leathery leaves. The inflorescence is pretty: magenta and white. An oddity that is best grown out of a pot on a piece of bark or a branch.

T. butzii. Small, with thin, twisted, cylindrical leaves that are purple spotted. The bracts are rose-colored with purple petals and yellow stamens. A pretty species that blooms in spring.

T. caput-medusae. Resembles *T. bulbosa.* Small, with vivid blue flowers.

74

T. cyanea. Grassy foliage and large purplish flowers. Outstanding [*page 71*].

T. flexuosa. Sometimes called *T. aloifolia.* Has coppery-green twisted leaves with silver crossbands. Bears red bracts and white flowers.

T. ionanthe. Dwarf, hardly more than 2 inches high. It will astound you at blooming time in spring when all the leaves blush fiery red and the pretty purple flower pokes its head through the foliage. Grow it in sun or bright light; needs little care [*page 71*].

T. juncea. Small and pretty, with narrow leaves in a tufted growth and a red flower crown. A favorite of mine.

T. punctulata. Narrow silver-gray pointed leaves. Bears a heavy inflorescence that is densely set with rose-red bracts and purple and white flowers. A good small air plant.

T. tricolor. Medium-sized, with grayish-green leaves edged red. The pink and red inflorescence is upright and branched, rising well above the plant.

Vriesea fenestralis. Regal plant, with green leaves delicately figured with darker green and purple lines. Big and bushy, but worth the space it takes; a most desirable bromeliad with yellow flowers. Grow this one warm.

V. malzinei. Rarely grows above 16 inches, with a compact rosette of claret-colored leaves shaded green. Fine for the windowsill. Early in spring the cylindrical flower head has yellow bracts with green margins. A very easy-to-grow plant.

V. splendens. Another good Vriesea. Grows about 12 inches high, and is a perfect houseplant. The green foliage is mahogany-striped; the thin thrusting spring and summer inflorescence is orange-colored.

Philodendron corcovadense (*photo by M. Barr*).

Hylocereus undatus.

Epiphyllum hybrid.

Platycerium ellisii
(*photo by J. Barnich*).

Platycerium bifurcatum
(*photo by J. Barnich*).

Philodendrons

These popular houseplants have been with us for years. Although they are essentially thought of as pot plants, many are epiphytic, and do very well on totem poles or pieces of wood. Indeed, the main consideration with vining philodendrons is that top leaves become smaller as the plant grows. If grown on poles or bark, where the large aerial roots can have something to grasp, leaves become larger as the plant grows.

Being mainly vining plants philodendrons need tall supports to be at their best. It is not necessary to hang plants, because they do fine in a suitable container; the support allows them to naturally vine as they would in nature. It is important, however, that the aerial roots (quite thick in many species) have something moist to grow into.

Most philodendrons have lance-shaped leaves. Some are complete, but others are cut leaf. Foliage color varies from golden-green to a dark lush green, and plants grow quickly in good light and humidity (sun is not necessary).

Philodendrons are basically foliage plants, so do not expect flowers in the home. Some of the philodendrons I have grown and you might want to try are:

P. andreanum.   Handsome, arrow-shaped foliage; needs moisture and warmth.

P. bipinnatifidum.   Large, with waxy, green, stiff, and segmented leaves to 24 inches. Climber.

P. corcovadense.   Dark green beautifully veined leaves [*page 76*].

P. cordatum (*P. oxycardium*).   The heart-leaf plant. Glossy green leaves. Grows in water or soil.

P. hastatum variegatum.   New, with yellow and green leaves [*opposite*].

P. imperialis.   Mammoth crinkled leaves; stellar plant [*opposite*].

Philodendron imperialis
(*photo by M. Barr*).

Philodendron
hastatum
(*photo by M. Barr*).

P. panduraeforme.   Scalloped olive green leaves; grows low.

P. pertusum.   Robust grower, with deep-lobed, heart-shaped leaves. Variegated form of yellow and green is also available.

P. soderoi.   Large or small forms, with mottled leaves and red stems.

P. squamiferum.   Unusual leaf design; good accent.

P. verrucosum. Outstanding. Exotic satin-sheen foliage. Needs warmth and humidity.

Ferns

We are all familiar with the lovely Boston ferns and their varieties. But the fern group includes some overlooked plants that are epiphytic and a joy to grow indoors. Most people have trouble with ferns such as Davallias, Platyceriums, Nephrolepis species, because they grow them in pots. These are epiphytic plants and thus prefer to be grown on slabs of wood or even on rocks. It takes them longer than expected to get established, but once growing they cover the support and make a beautiful picture.

Davallias (rabbit's foot ferns) simply do not grow well in a pot and so must be cultivated on slabs of wood or bark. Platyceriums are unusual plants with large staghorn-type fronds—giving them the name staghorn ferns—and healthy specimens are handsome indeed.

The Mexican tree fern is still another fern that can be successfully grown on wood or simply in gravel in a dish of water. Like the resurrection fern, this fern will wither in adverse conditions, but as soon as moisture and light are supplied, it revives and unfurls its beautiful fronds.

Here are some of my air-garden ferns that I have grown through the years:

Adiantum tenerum.   Elegant fern, with glossy green fronds and wiry black stems. Always handsome.

Asplenium bulbiferum. This favorite from New Zealand and Australia is popular because it grows rapidly and tolerates a dry atmosphere. The fronds are 24 inches long and about 10 inches broad. The plant has a rather upright habit, although the fronds are usually pendulous, covered with young plants at certain times of the year.

Cibotium schiedei. One of the most handsome ferns you can find, native to Mexico and Guatemala. Fronds are elegantly cascading, triangular in shape, and can grow to 10 feet in length. They are three-times divided and borne on stout brownish trunks from crowns covered with dense brown hairs. The spear-shaped leaflets come to a narrow point. A very lacy and ornamental plant.

Davallia bullata mariesii. Small, fine, lacy fronds; creeping plant with hairy rhizomes. Can be trained to grow on almost any support and assume the shape of the support.

D. fejeenisis plumosa. Charming fern from the Fiji islands. Has thick rhizomes and handsome, finely cut fronds that are borne on upright stalks and are 6 to 9 inches long and 12 inches broad. Fronds make a deltoid outline and are brilliant green; they are somewhat pendant when mature. A really elegant fern that requires shade.

D. trichomanoides. Also called *D. canariensis.* Wiry gray stalks with leathery 6- to 9-inch fronds. Durable, and the creeping rhizomes adapt to totem poles or osmunda slabs.

Humata tyermannii. Related to Davallias. Dark green, leathery, 6-inch fronds. Slow-growing and remains small; good for rocks.

Nephrolepis exalta (Boston fern). A large group of lovely green ferns with many varieties. Most people find these popular plants difficult to grow, but if in a porous air plant medium, such as shredded tree-fern or bark with some soil, plants thrive.

81

Davallia Fern.

Kevin Haapala '73

Platycerium bifurcatum majus. Broad rich green, pendant and leathery fronds. An easy-growing epiphyte that grows best on a large slab of wood [*page 77*].

P. coronarium. Large clustering fern that grows high in trees in the rain forests. The green fronds are pendulous, wide, and forked. Well-grown specimens are magnificent to see.

P. ellisii. Has a fresh green color, with rounded basal leaves; fertile fronds are erect, fan-shaped, and divided at the lobes [*page 77*].

P. vassei. A compact Platycerium with short, stiff, and up-right fronds. Good for a small slab of osmunda.

P. veitchii. A very vigorous tree dweller, with cupped basal, leathery, dark green, fertile, and forked fronds.

Polypodium aureum undulatum. Graceful and arching with bluish-green, deeply lobed fronds. Best grown in osmunda or on tree bark. (Rarely succeeds in soil.)

P. subauriculatum. Decorative basket fern, with graceful fronds borne on heavy rhizomes. Slow-growing but desirable.

Pteris cretica. An unusual plant, with somewhat feathery growth and many variants. The typical species comes from the Himalayas and has 6- to 12-inch fronds that are borne on erect wiry stalks. The leaves are somewhat leathery and have a papery texture. You will find varieties with crested, sword-shaped, and fan-shaped leaves.

P. ensiformis 'Victoriae'. Much prettier than the species, this fern is slender and graceful, with either short leaves, or tall, slender leaves. The leaves have sharply serrated edges and are marked with silver bands and edged in green. A totally handsome plant.

Cactus

It is hard to believe that the Christmas and Easter cacti, namely, Zygocactus and Rhipsalidopsis, respectively, grow as epiphytes in nature, often in the company of orchids. But they do, and beautifully, on supports rather than in pots. The flowers of these handsome cacti are breathtakingly beautiful in shades of pink, purple, or cerise, and at holiday times they are a halo of color.

Another cactus group that can be grown as epiphytes are Rhipsalis cacti, with their long cascading pencil-like leaves; there are several species you might want to try.

The basic requirements for all air plant cacti are the same as for orchids or bromeliads: a good supply of moisture at the roots with spraying, good humidity, and bright light.

Here are some of the best cacti to use as air plants:

Aporocactus flagelliformis (rattail cactus).   Do not let the name scare you; this is a handsome plant, with slender hanging stems covered with small brownish spines. The lovely flowers are tubular and cerise. This epiphyte can be grown on rocks or in osmunda baskets. Water heavily in summer, but not so much in winter.

Epiphyllum hybrids.   Pendant, flattened leaves with scalloped edges. Plants bear large flowers in July or August. There are dozens of varieties, one prettier than the other. Grow in natural tree supports for striking effect [*page 76*].

Hatiora salicornioides.   These plants are closely related to Rhipsalis. The leaves are branching, slender, and jointed, and bear greenish-yellow flowers. Unusual but lovely and very easy to grow in moss baskets.

Hylocereus undatus.   This Hylocereus has deep green triangular joints, with long climbing or drooping habit. Flowers are mammoth; some 12 inches across. Plants can grow huge and so need suitable large supports [*page 76*].

84

Rhipsalidopsis (schlumbergera) gaertneri (Easter cactus). Branching plant, with leaflike joints with serrated edges and small bristly spines. Plant bears red flowers. Lovely in hanging osmunda planters.

Rhipsalis cruciforme. Branching plant, with angular or flat leaflike stems cascading to 6 feet. Greenish flowers are tiny. Plant is easy to grow.

R. houlletiana. Branching, somewhat erect plant, with tall cylindrical stems. Bears small cream flowers.

R. paradoxa. Winged stems and twisting growth; unusual hanging plant.

R. rhombea. Dark green, scalloped, jointed leaves. Best grown in hanging containers.

Selenicereus grandiflorus. Straggly cactus that needs ample space. A trailing or climbing plant that wraps itself quickly around any support. Leaves are grayish-green, and the nocturnal fragrant white flowers are immense. Can grow very large.

Zygocactus truncatus. The Thanksgiving or crab cactus (but still referred to as the Christmas cactus). Claw-shaped dark green (chains); can grow into an impressive plant. Requires complete darkness for about 4 to 6 weeks before bloom time.

7·Other Plants to Try

Chapter 6 lists many of the true epiphytes and other candidates for air gardening; here we look at some plants that, although not true epiphytes and little known as air plants, will grow for some time on pieces of wood, stones, and so forth. These plants include many of the common houseplants and are certainly worthwhile additions to the indoor garden.

So if you are tired of the same old plant in a pot at the windowsill, try something different and make a dish garden or grow a plant anchored to a piece of wood. Create handsome stone landscapes in trays or dishes and tuck popular houseplants like *Acorus gramineus* or Peperomias in place.

One of the best plants that come to my mind (it is often sold as a plant that grows without soil) is the ti plant *Cordyline terminalis.* All this really needs is a dish of water to grow in. But such an arrangement is hardly pleasing, so it is much better to grow it in the same fashion as an air plant for a decorative accessory.

Other plants include the popular Sansevierias, Chinese evergreens, Pothos, Syngoniums, some succulents, ivy, and one of my favorites: *Chlorophytum elatum,* the spider plant. Here are the best plants to try:

Acorus gramineus pusillus.   Charming dwarf, grassy plant that is ideal for wood and slab growing. Does well in almost any situation, and the grassy leaves make it a vivid splash of indoor color.

Aglaonema commutatum (Chinese evergreen).   Popular houseplant, with deep and shiny dark green spear-shaped leaves. Another large plant, and a somewhat rapid grower. Prefers shady place [*page 89*].

Anthurium crystallinum.   An exotic plant with exquisite velvety green leaves veined white. *A. andreanum* also handsome [*page 90*].

Chlorophytum elatum (spider plant).   Nondestructible plant that has its own water-storage vessels and does well on a suitable wood support. Plants are grassy and green and grow in rosette form. Excellent [*page 90*].

Cibotium menzenii (Hawaiian tree fern).   This big and lacy fern comes as a rootstock from suppliers. Simply anchor the root in a suitable dish with stones; in no time, graceful fronds will appear.

Cordyline terminalis (ti plant).   Good plant for dish growing on stones and rocks. Leaves are dark green, with handsome red edges spirally arranged. "Tricolor" is a pink-, red-, and creamy white-leaved variety I find especially handsome.

Dracaena goldieana.   Beautiful rosette of green and yellow or white leaves. Grows very large, and its graceful shape makes it ideal for bark growing. *D. warneckii* is also handsome.

Echeveria.   This is a large group of plants with many new varieties. Basically, echeverias are pale green or gray-green in color, leaves pointed and generally in rosette growth. Plants appear like carved jade and while these are not air plants, but

Scindapsus aureus (*Merry Gardens photo*).

Dracaena 'Perfection' (*photo courtesy Alberts & Merkel*).

Dracaena goldieana (*photo courtesy Alberts & Merkel*).

*Peperomias in a moss basket
(photo by author).*

*Aglaonemas
(Merry Gardens photo).*

Anthurium andraeanum (*left*) *and* A. crystallinum
(*right*) (*photo by U.S.D.A.*).

Chlorophytum growing in a rock (photo by author).

rather succulents, they can certainly be tried for wall panel
growing or in sphagnum moss in natural supports of wood or
stone.

Hedera helix (ivy).   There are dozens of varieties of this favorite green plant. Generally leaves are scalloped and pointed.
Some varieties, and the best for air plant growing, are the very
small-leaved ones like 'Curlilocks' and 'Green Imp'.

Cordyline Terminalis.

Peperomia. This large group of charming small plants contains a few true epiphytes and easier plants to grow are impossible to find. *Peperomia glabella* has fleshy-oblong glossy leaves and grows in a somewhat twining manner. *P. obtusifolia* has dark green shiny leaves and is more bushy. Both plants do fine on small chunks of rotted tree bark [*page 89*].

Pothos (*scindapsus*) *aureus* (devil's ivy). This amenable vine grows easily. It has broad green leaves blotched with yellow. Easy to grow and always good. Try it tucked between rocks in a dish.

Syngonium podophyllum (arrowhead). These are climbing vines with shiny green leaves shaped like arrowheads. Plants grow rapidly and to a large size, so use suitable large supports for them. *S. wendlandii* is a fine variegated form that is very handsome.

Sedums. A large group of succulent plants, many small ones excellent for air plant growing. *S. pachyphyllum* and *S. stahlii* easy to grow.

Tradescantia. These viners are absolutely fine for air gardening as companions to bromeliads or orchids or wherever you need some good color accent. There are many species; two I work with are *T. blossfeldiana* and *T. fluminensis.*

These are some of the houseplants I have grown successfully in air plant gardens, but the list is by no means complete. You can certainly try other plants to see if they succeed in this type culture. Half the fun of any kind of gardening is enjoyment and satisfaction, and growing air plants or any plant in a new way is exciting as well.

Where to Buy Air Plants

There are many plant suppliers; some specialize in ferns, others in orchids, and still others in cacti. Look for suppliers in the classified sections of garden magazines, under specific plant categories. Send for lists and catalogs first to decide just what you want and how to arrange your air gardens: trellis, on rocks, in tree branches, and so forth. The size of the object you intend growing the plant on will dictate the size of the plant. Select accordingly. Plants are nominally priced in the small- to medium-sized category, and shipment is generally fast.

Order in spring and fall. In spring, good weather is on the way, and there is little chance of plants dying from overheating; in fall there is still moderate weather. Avoid summer buying because heat can kill plants quickly.

When you receive plants, even from reliable nurseries, they may seem somewhat wan, but do not fret. Once you have them established on their supports, fresh green growth starts as you increase moisture.

Some air plants (such as ferns and cacti) can be found at local nurseries or plant stores. However miniature orchids

Bromeliad.

Kevin Haapal

and small bromeliads are generally only available through mail-order suppliers; here is a partial list of such companies:

Alberts & Merkel Bros., Inc.
P.O. Box 537
Boynton Beach, Fla. 33435

Orchids, bromeliads, philodendrons; catalog charge.

Hauserman's Orchids
Box 363
Elmhurst, Ill. 60126

Orchids; many miniatures; catalog free.

Henrietta's Nursery
1345 N. Brawley Ave.
Fresno, Calif. 93705

All kinds of cacti; catalog charge.

Logee's Greenhouses
55 North St.
Danielson, Conn. 06239

Small ferns and other miniature plants. Catalog charge.

Margaret Ilgenfritz Orchids
P.O. Box 665
Monroe, Mich. 48161

Many miniature orchids; catalog charge.

Merry Gardens
Camden, Me. 04843

Small ferns, philodendrons, and other miniature plants. Catalog charge.

Oak Hill Gardens
P.O. Box 25
W. Dundee, Ill. 60118

Fine stock of bromeliads, orchids. Pamphlet, free.

Seaborn Del Dios Nursery
1220 41st. Ave.
Escondido, Calif. 92025

Large selection of bromeliads; catalog charge.

Walther's Exotic House Plants
Dept. H, R.D. #3,
9-W Highway
Catskill, N.Y. 12414

Tillandsias are their specialty.

Where to Buy Supplies

Totem poles and macrame or basket slings are at houseplant shops and nurseries more and more; so are trellises, shallow dishes for dish gardening, wood slabs and plaques—availability depending on the size of the supplier.

Sphagnum moss and osmunda are generally difficult to find, but they can be ordered by mail from most orchid suppliers. In their catalogs you will find these items advertised. Pressed Hawaiian tree-fern slabs are also available as well as tree-fern pots and baskets.

First stop for wire for attaching plants is at your hardware dealer, though you can sometimes buy florist wire from local florist shops.